Calculators

number patterns and Magic

BURELOFF
JOHNSON
ROES

FOREWORD

Dear Fellow Teachers,

There are many calculator books. There are books about the care and feeding of calculators, books about the consumer uses of calculators, textbooks based on the use of calculators, books of calculator games and curiosities. This is yet another calculator book but just a little different from the others. These lessons were developed to be used with a calculator, not to teach calculator skills and certainly not to be used as non-calculator busy work. We believe that these lessons present the calculator as an aid in the discovery aspects of number patterns and as a device for checking predictions made about number patterns.

Morris Bureloff contributed much of the original manuscript, Ruth Roes did the intriguing artwork and I made additions and deletions, organized and edited. Mary Laycock of Activity Resources put it, and us, together. The four of us sincerely hope that you and your students will enjoy adding a new dimension to your calculator experiences.

July, 1977 — Connie Johnson

Activity Resources Company, Inc.
P. O. Box 4875
Hayward, California 94540

TABLE OF CONTENTS

Magic* Squares 1

name ______________________

period ______________________

date ______________________

SQUARE # 1

Use the digits 1-9 to make a magic square.

SQUARE # 2

Multiply each entry in SQUARE # 1 by 9 to form a new square. Do you still have a MAGIC SQUARE?

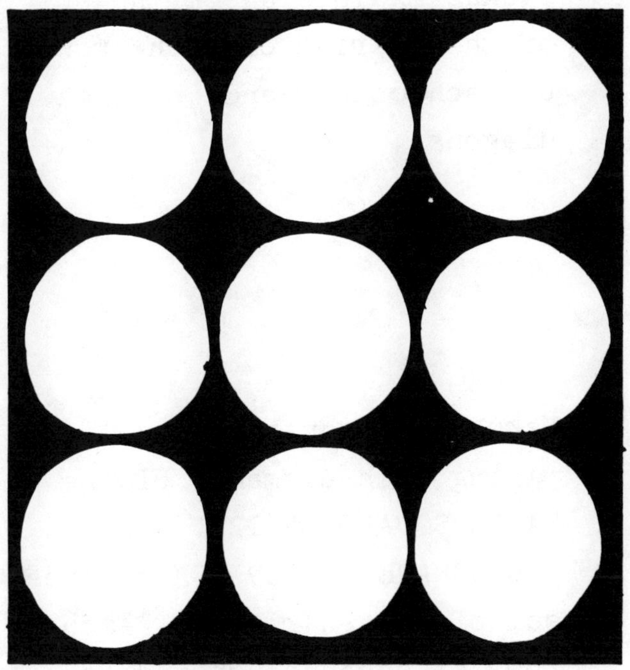

SQUARE # 3

Add 2 to each entry of SQUARE # 1 to form a new square. Is the new square still MAGIC?

Magic* Squares 2

name ______________________

period ______________________

date ______________________

This is a MAGIC SQUARE.
Find the sum of the entries of each row, of each column and of each diagonal.

	2	7	6	
	9	5	1	
	4	3	8	

Make a new square by adding 765 to each element of the MAGIC SQUARE. Is the new SQUARE MAGIC? Check the sum of the entries of each row, of each column and of each diagonal.

Make a new square by multiplying each element of the MAGIC SQUARE by 123. Is the new SQUARE MAGIC? Check the sum of the entries of each row, of each column and of each diagonal.

Magic* Squares 3.

name ____________________

period ____________________

date ____________________

Complete this MAGIC SQUARE. It is built from the numbers 1-25.

17				15
	5		14	
		13		
	12		21	
11				9

Magic* Squares 4.

name ______________

period ______________

date ______________

Complete this MAGIC SQUARE. It is built from the numbers 1-49.

	38		1		19	
	46					
					35	
5		16				
			33		43	
21		32		49		12
		39			11	

MULTIPLICATIVE Magic* Squares 1.

name ______________________

period ______________________

date ______________________

Given the numbers 1, 2, 3, 4, 6, 9, 12, 18, 36. Arrange them to form a multiplicative MAGIC SQUARE.*** The product of each row, column and diagonal is the same.

Write the numbers in the MAGIC SQUARE in prime factored form.

MULTIPLICATIVE * Magic Squares 2.

name ______________

period ______________

date ______________

Use the numbers 1, 3, 5, 9, 15, 25, 45, 75, 225 to form a multiplicative MAGIC SQUARE.

Write the numbers in the MAGIC SQUARE in prime factored form.

Compare the prime factor form of MULTIPLICATIVE MAGIC SQUARE - 1 with the prime factor form of MULTIPLICATIVE MAGIC SQUARE - 2. Do you see the pattern for making MULTIPLICATIVE MAGIC SQUARES?

MULTIPLICATIVE Magic* Squares 3.

name ______________________

period ______________________

date ______________________

You will need to use the pattern for MULTIPLICATIVE MAGIC SQUARES that you discovered from the prime factor patterns.

Write a multiplicative magic square using the numbers 1, 2, 4, 5, ____, ____, ____, ____, ____

Write a MULTIPLICATIVE MAGIC SQUARE using the numbers 1, 2, 4, 7, ___, ___, ___, ___, ___

MULTIPLICATIVE Magic* Squares 4.

Name ____________

period ____________

date ____________

Write a MULTIPLICATIVE MAGIC SQUARE using the numbers
1, 3, 7, ___, ___, ___, ___, ___, ___.
Predict the MAGIC PRODUCT.

Write a MULTIPLICATIVE MAGIC SQUARE using the numbers
1, 3, 9, 11, ___, ___, ___, ___, ___.
Predict the MAGIC PRODUCT.

Magic* Wheels and Triangles*

name ____________________

period ____________________

date ____________________

For each of the three figures, arrange the digits 1-9 so that the sum along each row of numbers is the same.

Magic Stars

name ____________________

period ____________________

date ____________________

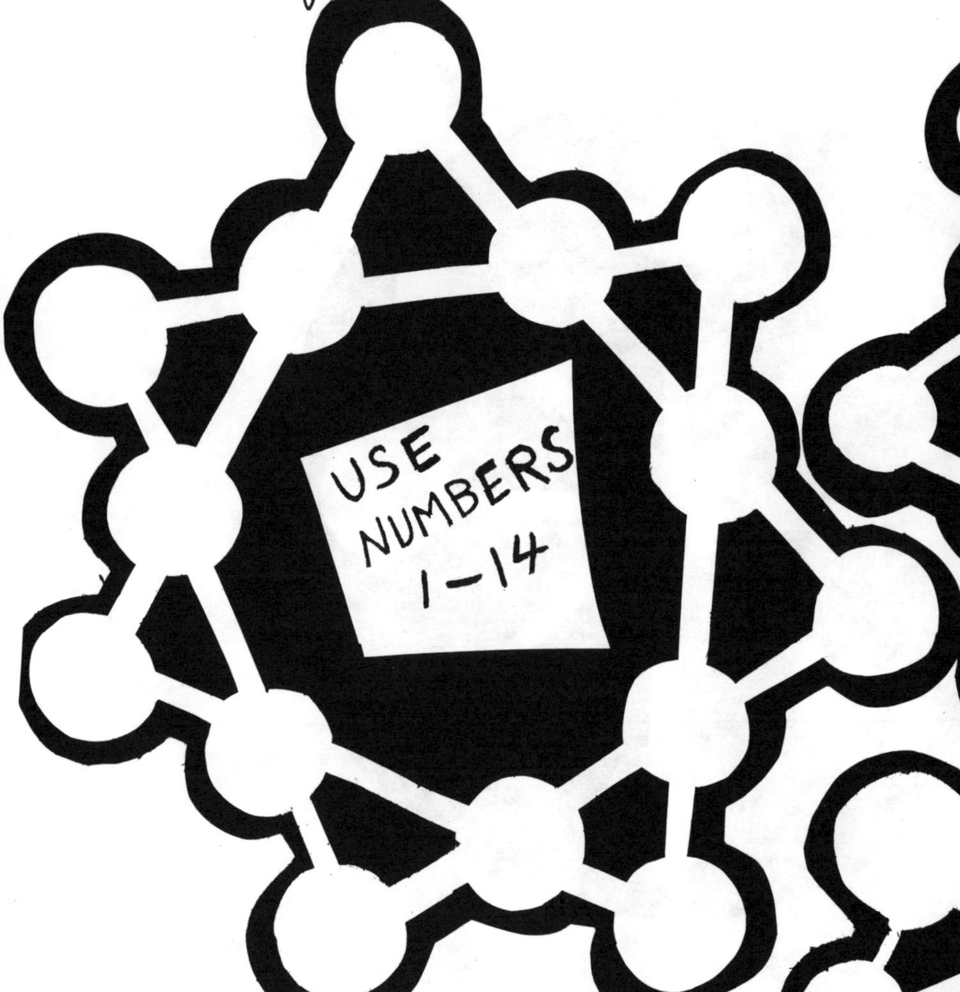

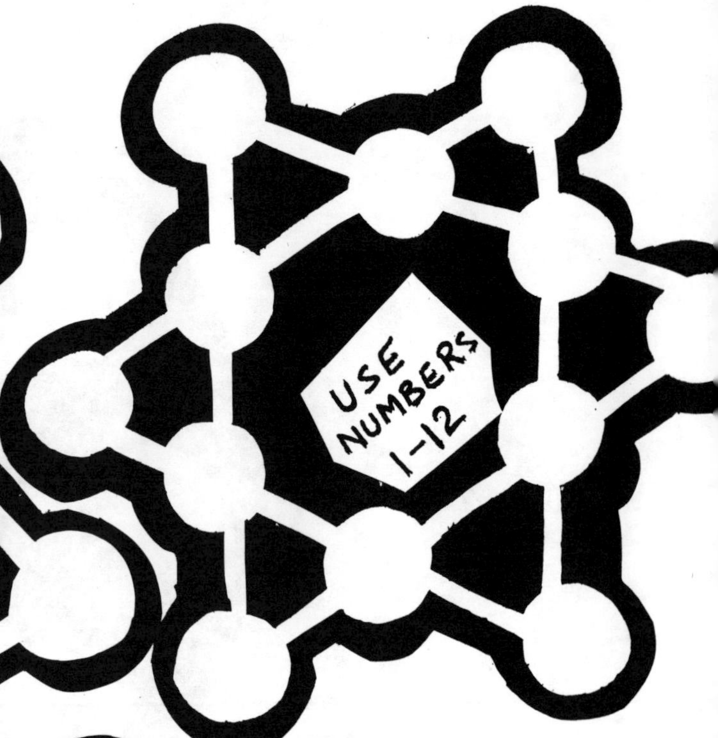

Arrange the given numbers so that the sums along each row are the same.

Magic Maltese Multiplication

name ____________________
period ____________________
date ____________________

Cut out the given number circles. Arrange the circles on the diagram so the product of any three numbers in line across the center or the product of any three numbers on the edge of an arm are the same.

1 2 3 4 6 9 12 18 36

2 3 12 18

Magic Roman Addition Cross

name ______________________

period ______________________

date ______________________

Cut out the given number circles. Arrange the numbers on the diagram so you can add any three or five numbers in line across the center of the cross or any four at the end of an arm to get the same sum. Turn the page upside down and find the sums on the new cross. Only two sums will vary. Find them.

- -

II XIX XX X+I III XX X+I II

I XI I I IX I XIX I

Magic* Patterns 1.

name
period
date

Find the quotients. Write the answer that begins with 1 in answer blank 1, the answer that begins with 2 in answer blank 2, etc. Write the letter at the end of the answer blank.

a. $\frac{2}{9} =$	b. $\frac{9}{9} \times 9 =$	c. $\frac{7}{9} =$
d. $\frac{1}{9} =$	e. $\frac{2}{3} =$	f. $\frac{8}{9} =$
g. $\frac{5}{9} =$	h. $\frac{1}{3} =$	i. $\frac{4}{9} =$

1	4	7
2	5	8
3	6	9

Magic* Patterns 2.

name ________________
period ________________
date ________________

Find the products. Write the answer that begins with 1 in answer blank 1, the answer that begins with 2 in answer blank 2, etc. Write the letter at the end of the answer in the blank.

a. 12,345,679 × 81

b. 12,345,679 × 63

c. 12,345,679 × 36

d. 12,345,679 × 72

e. 12,345,67 × 9

f. 12,345,679 × 27

g. 12,345,679 × 54

h. 12,345,679 × 18

i. 12,345,679 × 45

1.	4.	7.
2.	5.	8.
3.	6.	9.

Magic* Patterns 3.

name
period
date

Find the products. Write the answer that begins with 1 in answer blank 1, the answer that begins with 2 in answer blank 2, etc. Write the letter at the end of the answer blank.

a. 10,101 × 77

b. 3,367 × 132

c. 8,547 × 78

d. 37,037 × 27

e. 1,443 × 77

f. 2,849 × 78

g. 15,873 × 56

h. 5,291 × 105

i. 3,003 × 111

1.

2.

3.

4.

5.

6.

7.

8.

9.

Magic Patterns 4.

name ______
period ______
date ______

Find the products. Write the answer that begins with 1 in answer blank 1, the answer that begins with 2 in answer blank 2, etc. Write the letter at the end of the answer in the blank.

a. 37×9

b. 37×27

c. 37×3

d. 37×18

e. 37×6

f. 37×15

g. 37×21

h. 37×12

i. 37×24

1.	4.	7.
2.	5.	8.
3.	6.	9.

Magic* Patterns 5.

name

period

date

Find the quotients. Write the answer that begins with 1 in answer blank 1, the answer that begins with 2 in answer blank 2, etc. Write the letter at the end of the answer in the blank.

a. $\frac{4}{333} =$	b. $\frac{33}{37} =$	c. $\frac{41}{333} =$
d. $\frac{263}{333} =$	e. $\frac{26}{111} =$	f. $\frac{226}{333} =$
g. $\frac{21}{37} =$	h. $\frac{115}{333} =$	i. $\frac{152}{333} =$

0	3	6
1	4	7
2	5	8

Magic* Powers 1.

name ____________________

period ____________________

date ____________________

$1^2 + 3^2 = \square\square$

$8^2 + 6^2 = \square\square\square$

$18^2 + 26^2 = \square\square\square\square$

$28^2 + 96^2 = \square\square\square\square\square$

$12^2 + 316^2 = \square\square\square\square\square\square$

$352^2 + 936^2 = \square\square\square\square\square\square\square$

Magic* Powers 2.

name ______________________

period ______________________

date ______________________

$1^3 + 5^3 + 3^3 = \square\square\square$

$3^3 + 7^3 + 0^3 = \square\square\square$

$3^3 + 7^3 + 1^3 = \square\square\square$

$4^3 + 0^3 + 7^3 = \square\square\square$

$3^3 + 4^3 + 5^3 = \square^3$

$1^3 + 6^3 + 8^3 = \square^3$

$6^3 + 8^3 + 10^3 = \square\square^3$

$12^3 + 1^3 - 9^3 = \square\square^3$

Magic* Powers 3.

name ______________________

period ______________________

date ______________________

$3^2 + 4^2 = \square^2$

$5^2 + 12^2 = \square\square^2$

$8^2 + 15^2 = \square\square^2$

$7^2 + 24^2 = \square\square^2$

$20^2 + 21^2 = \square\square^2$

$12^2 + 35^2 = \square\square^2$

$40^2 + 9^2 = \square\square^2$

$28^2 + 45^2 = \square\square^2$

$11^2 + 60^2 = \square\square^2$

$33^2 + 56^2 = \square\square^2$

$16^2 + 63^2 = \square\square^2$

$48^2 + 55^2 = \square\square^2$

$36^2 + 77^2 = \square\square^2$

$13^2 + 84^2 = \square\square^2$

$39^2 + 80^2 = \square\square^2$

$65^2 + 72^2 = \square\square^2$

Magic* Powers 4.

name ______________________

period ______________________

date ______________________

$\sqrt{81}$ =	8 + 1 =	$(\quad)^2 = 81$
$\sqrt{2025}$ =	20 + 25 =	$(\quad)^2 = 2025$
$\sqrt{3025}$ =	30 + 25 =	$(\quad)^2 = 3025$
$\sqrt{9801}$ =	98 + 01 =	$(\quad)^2 = 9801$
$\sqrt{88209}$ =	+ =	$(\quad)^2 = 88209$
$\sqrt{494209}$ =	+ =	$(\quad)^2 = 494209$
$\sqrt{998001}$ =	+ =	$(\quad)^2 = 998001$

Magic* Operations* 1.

name ____________________

period ____________________

date ____________________

IS MULTIPLICATION OF FRACTIONS THE SAME AS SUBTRACTION

$1 \times \frac{1}{2} =$ ______	$1 - \frac{1}{2} =$ ______
$2 \times \frac{2}{3} =$ ______	$2 - \frac{2}{3} =$ ______
$3 \times \frac{3}{4} =$ ______	$3 - \frac{3}{4} =$ ______
$4 \times \frac{4}{5} =$ ______	$4 - \frac{4}{5} =$ ______
$5 \times \frac{5}{6} =$ ______	$5 - \frac{5}{6} =$ ______
$6 \times \frac{6}{7} =$ ______	$6 - \frac{6}{7} =$ ______
$7 \times \frac{7}{8} =$ ______	$7 - \frac{7}{8} =$ ______

GIVE ONE (ONLY ONE) COUNTER-EXAMPLE.

Magic* Operations* 2.

name ____________________

period ____________________

date ____________________

IS DIVISION OF FRACTIONS THE SAME AS ADDITION OF FRACTIONS

$1\frac{1}{3}+\frac{2}{3}=$ ______	$1\frac{1}{3}\div\frac{2}{3}=$ ______
$2\frac{1}{4}+\frac{3}{4}=$ ______	$2\frac{1}{4}\div\frac{3}{4}=$ ______
$3\frac{1}{5}+\frac{4}{5}=$ ______	$3\frac{1}{5}\div\frac{4}{5}=$ ______
$4\frac{1}{6}+\frac{5}{6}=$ ______	$4\frac{1}{6}\div\frac{5}{6}=$ ______
$5\frac{1}{7}+\frac{6}{7}=$ ______	$5\frac{1}{7}\div\frac{6}{7}=$ ______
$6\frac{1}{8}+\frac{7}{8}=$ ______	$6\frac{1}{8}\div\frac{7}{8}=$ ______
$7\frac{1}{9}+\frac{8}{9}=$ ______	$7\frac{1}{9}\div\frac{8}{9}=$ ______

GIVE ONE (ONLY ONE) COUNTER-EXAMPLE

Magic* Operations* 3.

name ____________________

period ____________________

date ____________________

IS MULTIPLICATION OF FRACTIONS THE SAME AS ADDITION OF FRACTIONS?

$1\frac{1}{2} \times 3 =$ ____	$1\frac{1}{2} + 3 =$ ____
$1\frac{1}{3} \times 4 =$ ____	$1\frac{1}{3} + 4 =$ ____
$1\frac{1}{4} \times 5 =$ ____	$1\frac{1}{4} + 5 =$ ____
$1\frac{1}{5} \times 6 =$ ____	$1\frac{1}{5} + 6 =$ ____
$1\frac{1}{6} \times 7 =$ ____	$1\frac{1}{6} + 7 =$ ____
$1\frac{1}{7} \times 8 =$ ____	$1\frac{1}{7} + 8 =$ ____
$1\frac{1}{8} \times 9 =$ ____	$1\frac{1}{8} + 9 =$ ____

GIVE ONE (ONLY ONE) COUNTER-EXAMPLE.

Magic* Operations* 4.

name ____________________

period ____________________

date ____________________

IS DIVISION OF FRACTIONS THE SAME AS SUBTRACTION OF FRACTIONS?

$4\frac{1}{2} \div 3 =$ ____	$4\frac{1}{2} - 3 =$ ____
$5\frac{1}{3} \div 4 =$ ____	$5\frac{1}{3} - 4 =$ ____
$6\frac{1}{4} \div 5 =$ ____	$6\frac{1}{4} - 5 =$ ____
$7\frac{1}{5} \div 6 =$ ____	$7\frac{1}{5} - 6 =$ ____
$8\frac{1}{6} \div 7 =$ ____	$8\frac{1}{6} - 7 =$ ____
$9\frac{1}{7} \div 8 =$ ____	$9\frac{1}{7} - 8 =$ ____

GIVE ONE (ONLY ONE) COUNTER-EXAMPLE.

Magic Matrix 1.

name ____________________

period ____________________

date ____________________

3	5	8	10
5	7	10	12
6	8	11	13
8	10	13	15

magic sum = ___

1. Circle any number in the first row of the matrix. Cross out all other numbers in the same row and all other numbers in the same column.
2. Circle any number in the second row which has not been crossed out. Cross out all other numbers in the same row and in the same column which have not yet been crossed out.
3. Circle any number in the third row which has not been crossed out. Cross out all other numbers in the same row and in the same column not yet crossed out.
4. Circle the remaining number.
5. Find the sum of the five circled numbers. The MAGIC SUM is ___.

3	5	8	10
5	7	10	12
6	8	11	13
8	10	13	15

magic sum = ___

6. Repeat the process with this matrix but choose different numbers to circle. The MAGIC SUM is _____.

Magic* Matrix* 1.

Page 2

name ____________________

period ____________________

date ____________________

3	5	8	10
5	7	10	12
6	8	11	13
8	10	13	15

magic sum = _____

7. Repeat the number choosing process but choose still different numbers. The MAGIC SUM is _____.

3	5	8	10
5	7	10	12
6	8	11	13
8	10	13	15

magic sum = _____

8. And just once more, choose and circle other numbers. The MAGIC SUM this time is _____.

9. What is the secret of the MAGIC MATRIX?

Magic Matrix 2.

name ______________________

period ______________________

date ______________________

-2	4	11	9	3
-10	-4	3	1	-5
-3	3	10	8	2
0	6	13	11	5
-20	-14	-7	-9	-15

magic sum= _____

1. Circle any number in the first row of the matrix. Cross out all other numbers in the same row and all other numbers in the same column.
2. Circle any number in the second row which has not been crossed out. Cross out all other numbers in the same row and in the same column.
3. Repeat the process for rows three, four and five. Five circled numbers should remain.
4. Find the sum of the five circled numbers. Write the MAGIC SUM.
5. Repeat the process with the matrix given below but choose five different numbers and find their sum. Write the MAGIC SUM.

magic sum= _____

-2	4	11	9	3
-10	-4	3	1	-5
-3	3	10	8	2
0	6	13	11	5
-20	-14	-7	-9	-15

Magic * Matrix * 2.

Page 2

name ______

period ______

date ______

-2	4	11	9	3
-10	-4	3	1	-5
-3	3	10	8	2
0	6	13	11	5
-20	-14	-7	-9	-15

magic sum = ______

6. Repeat the number choosing process using different numbers than those chosen before. Write the MAGIC SUM.

-2	4	11	9	3
-10	-4	3	1	-5
-3	3	10	8	2
0	6	13	11	5
-20	-14	-7	-9	-15

magic sum = ______

7. Use this matrix to check one more set of five numbers. Write the MAGIC SUM.

Magic* Matrix* 3.

name ______________________

period ______________________

date ______________________

The construction of a MAGIC MATRIX is simple. Do these steps.

1. Write a set of four numbers whose sum is 24. {3, 5, 6, 10}
2. Write another set of four numbers with sum 24. {2, 4, 5, 13}
3. Write each set of numbers along the side of a 4 x 4 square.
4. Write the sums of the numbers in the rows and columns.

	3	5	6	10
2	5			
4		9		
5			11	
13				23

magic sum = ______

Copy your MAGIC MATRIX and do the number-choosing process to find the MAGIC SUM. What is it? Check your findings.

Magic* Matrix* 4.

name ______________

period ______________

date ______________

Use the trick you learned in MAGIC MATRIX-3 to construct a 25-number MAGIC MATRIX. Check your matrix carefully and then try it on a friend.

Set 1 _____, _____, _____, _____, _____ SUM _____

Set 2 _____, _____, _____, _____, _____ SUM _____

name ________________

period ________________

date ________________

1.

Magic Mirror

= 9 X 9	9 + 9 =
= 2 4 X 3	3 + 2 4 =
= 4 7 + 2	2 X 4 7 =
= 4 9 7 + 2	2 X 4 9 7 =

Magic Mirror

= 1 2 X 1 2	2 1 X 2 1 =
= 1 3 X 1 3	3 1 X 3 1 =
= 1 0 2 X 1 0 2	2 0 1 X 2 0 1 =
= 1 0 3 X 1 0 3	3 0 1 X 3 0 1 =
= 1 1 2 X 1 1 2	2 1 1 X 2 1 1 =
= 1 1 3 X 1 1 3	3 1 1 X 3 1 1 =
= 1 2 2 X 1 2 2	2 2 1 X 2 2 1 =

name ______________________

period ______________________

date ______________________

2. Magic Mirror Magic

015	X 34 =	
018	X 45 =	
045	X 12 =	
0198	X 45 =	
0495	X 12 =	
0036	X 175 =	
10989	X 9 =	
21978	X 4 =	
04995	X 12 =	
01998	X 45 =	
00225	X 232 =	
00396	X 175 =	

name ______________________

period ______________________

date ______________________

3. Magic Mirror Mirror Magic

$= (1\,1\,1\,2)^2 \qquad (2\,1\,1\,1)^2 =$

$= (1\,1\,1\,3)^2 \qquad (3\,1\,1\,1)^2 =$

$= (1\,1\,2\,1)^2 \qquad (1\,2\,1\,1)^2 =$

$= (1\,1\,2\,2)^2 \qquad (2\,2\,1\,1)^2 =$

$= (1\,2\,1\,2)^2 \qquad (2\,1\,2\,1)^2 =$

$= (1\,1\,1\,1\,2)^2 \qquad (2\,1\,1\,1\,1)^2 =$

$= (1\,1\,1\,1\,3)^2 \qquad (3\,1\,1\,1\,1)^2 =$

$= (1\,1\,1\,2\,1)^2 \qquad (1\,2\,1\,1\,1)^2 =$

$= (1\,1\,1\,2\,2)^2 \qquad (2\,2\,1\,1\,1)^2 =$

$= (1\,1\,1\,1\,1\,2)^2 \qquad (2\,1\,1\,1\,1\,1)^2 =$

$= (1\,1\,1\,1\,2\,1)^2 \qquad (1\,2\,1\,1\,1\,1)^2 =$

$= (1\,1\,1\,2\,1\,1)^2 \qquad (1\,1\,2\,1\,1\,1)^2 =$

Magic* Numbers 1.

name ____________________

period ____________________

date ____________________

9109 X 1 = 0

9109 X 2 =

9109 X 3 =

9109 X 4 =

9109 X 5 =

9109 X 6 =

9109 X 7 =

9109 X 8 =

9109 X 9 =

9109 X 10 =

Magic* Numbers 2.

name ______________________

period ______________________

date ______________________

9	9	9	9	9	9	X		2	=							
9	9	9	9	9	9	X		3	=							
9	9	9	9	9	9	X		4	=							
9	9	9	9	9	9	X		5	=							
9	9	9	9	9	9	X		6	=							
9	9	9	9	9	9	X		7	=							
9	9	9	9	9	9	X		8	=							
9	9	9	9	9	9	X		9	=							
9	9	9	9	9	9	X	1	0	=							

Magic* Numbers 3.

name ____________________

period ____________________

date ____________________

You will find all digits from 1 to 9 in the problem and answer.

567 X 567 = ______

854 X 854 = ______

4761 X 069 = ______

57321 X 57321 = __________

60984 X 60984 = __________

16583742 X 90 = __________

32547891 X 60 = __________

51249876 X 30 = __________

303840 X 30384 = __________

Magic* Numbers 4.

name ______________________

period ______________________

date ______________________

111111111 X 9 = ☐☐☐☐☐☐☐☐☐

222222222 X 9 = ☐☐☐☐☐☐☐☐☐☐

333333333 X 9 = ☐☐☐☐☐☐☐☐☐☐

444444444 X 9 = ☐☐☐☐☐☐☐☐☐☐

555555555 X 9 = ☐☐☐☐☐☐☐☐☐☐

666666666 X 9 = ☐☐☐☐☐☐☐☐☐☐

777777777 X 9 = ☐☐☐☐☐☐☐☐☐☐

888888888 X 9 = ☐☐☐☐☐☐☐☐☐☐

999999999 X 9 = ☐☐☐☐☐☐☐☐☐☐

1111111110 X 9 = ☐☐☐☐☐☐☐☐☐☐

Magic* Numbers 5.

name ______________________

period ______________________

date ______________________

3	7	7	9	×	2	+	1	=				
2	5	1	9	×	3	+	2	=				
1	8	8	9	×	4	+	3	=				
1	5	1	1	×	5	+	4	=				
1	2	5	9	×	6	+	5	=				
1	0	7	9	×	7	+	6	=				
	9	4	4	×	8	+	7	=				
	8	3	9	×	9	+	8	=				
	7	5	5	×	10	+	9	=				

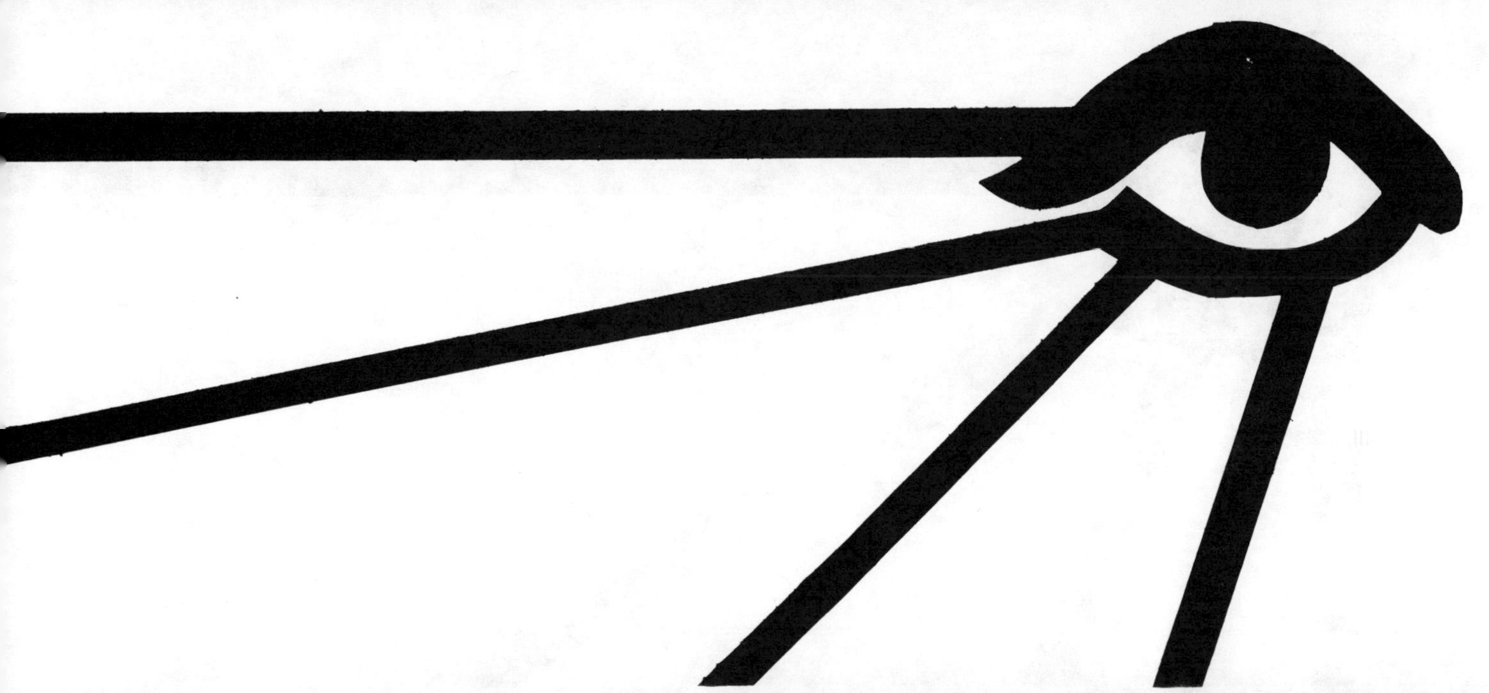

Magic* Numbers 6.

name ______________________

period ______________________

date ______________________

37037 X 3 =	
37037 X 6 =	
37037 X 9 =	
37037 X 12 =	
37037 X 15 =	
37037 X 18 =	
37037 X 21 =	
37037 X 24 =	
37037 X 27 =	
37037 X 30 =	
37037 X 33 =	
37037 X 36 =	
37037 X 39 =	
37037 X 42 =	
37037 X 45 =	
37037 X 48 =	
37037 X 51 =	
37037 X 54 =	

Magic* Numbers 7.

name ______________________

period ______________________

date ______________________

	1	5	8	7	3	×	7	=							
	3	1	7	4	6	×	7	=							
	4	7	6	1	9	×	7	=							
	6	3	4	9	2	×	7	=							
	7	9	3	6	5	×	7	=							
	9	5	2	3	8	×	7	=							
1	1	1	1	1	1	×	7	=							
1	2	6	9	8	4	×	7	=							
1	4	2	8	5	7	×	7	=							

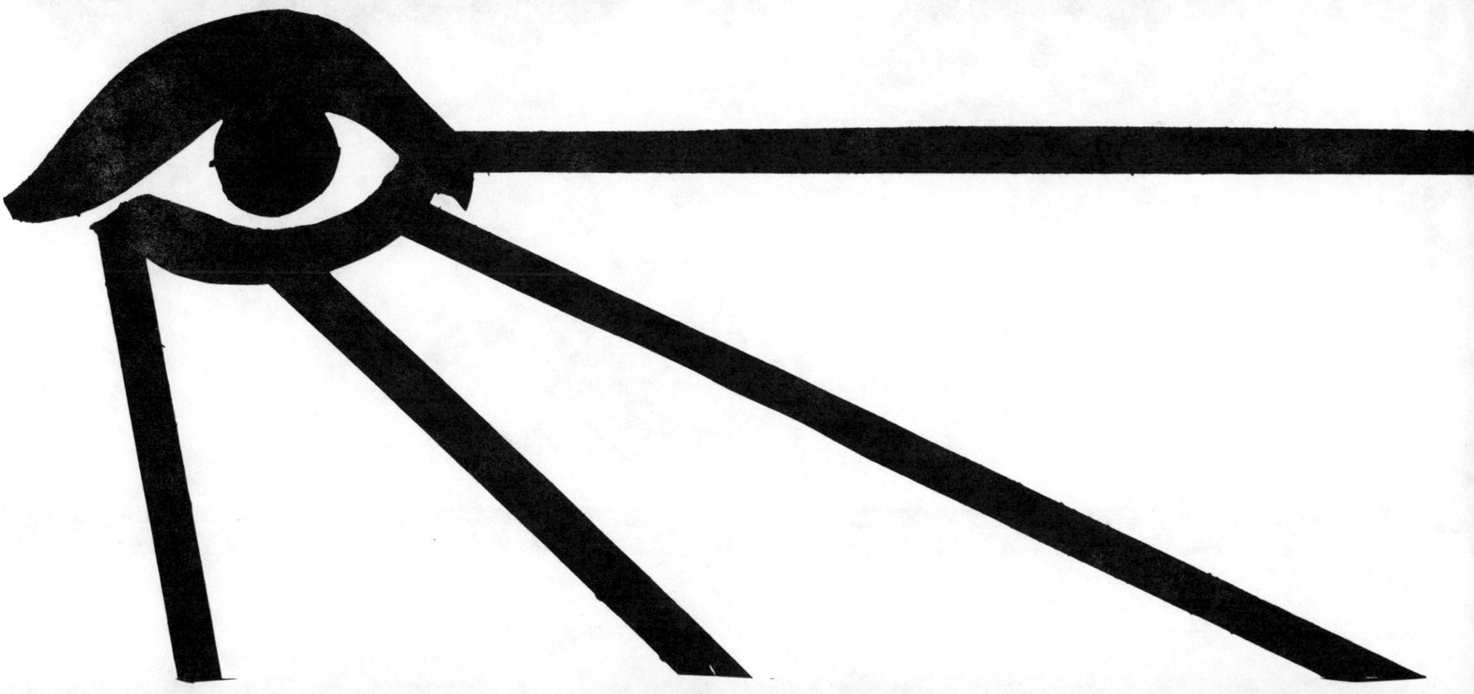

Magic* Numbers 8.

name ______________________

period ______________________

date ______________________

4	8	3	X	1	2	=				
1	3	8	X	4	2	=				
2	9	7	X	1	8	=				
1	9	8	X	2	7	=				
1	8	6	X	3	9	=				
1	5	9	X	4	8	=				
1	5	7	X	2	8	=				
1	7	3	8	X	4	=				
1	9	6	3	X	4	=				

BE CAREFUL! THERE IS NO "PATTERN" BUT EACH PROBLEM WITH ITS ANSWER CONTAINS THE DIGITS 1 — 9 ONCE AND ONLY ONCE.

Magic* Numbers 9.

name ____________________

period ____________________

date ____________________

3367 X 33 =	
3367 X 66 =	
3367 X 99 =	
3367 X 132 =	
3367 X 165 =	
3367 X 198 =	
3367 X 231 =	
3367 X 264 =	
3367 X 297 =	
3367 X 330 =	
3367 X 363 =	
3367 X 396 =	
3367 X 429 =	
3367 X 462 =	
3367 X 495 =	
3367 X 528 =	
3367 X 561 =	
3367 X 594 =	
3367 X 627 =	
3367 X 660 =	
3367 X 693 =	
3367 X 726 =	
3367 X 759 =	
3367 X 792 =	
3367 X 825 =	
3367 X 858 =	
3367 X 891 =	
3367 X 924 =	
3367 X 957 =	
3367 X 990 =	

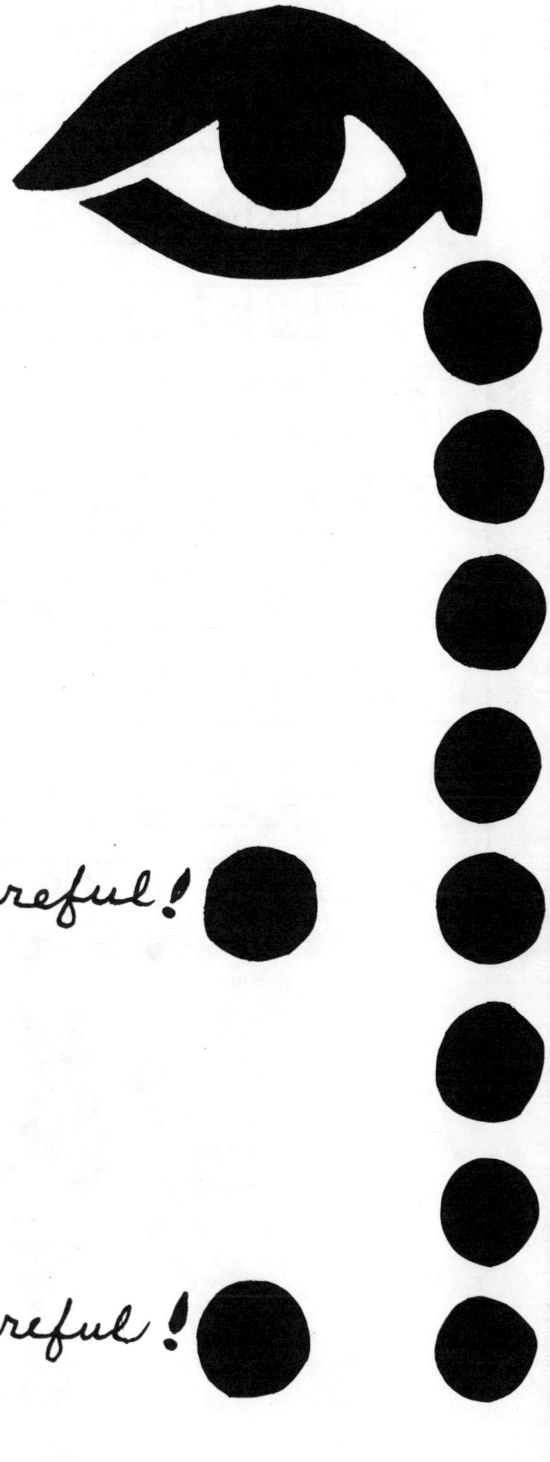

careful!

careful!

Magic* Numbers 10.

name ______________________

period ______________________

date ______________________

1 2 3 4 5 6 7 9 X 9 =	
1 2 3 4 5 6 7 9 X 1 8 =	
1 2 3 4 5 6 7 9 X 2 7 =	
1 2 3 4 5 6 7 9 X 3 6 =	
1 2 3 4 5 6 7 9 X 4 5 =	
1 2 3 4 5 6 7 9 X 5 4 =	
1 2 3 4 5 6 7 9 X 6 3 =	
1 2 3 4 5 6 7 9 X 7 2 =	
1 2 3 4 5 6 7 9 X 8 1 =	
1 2 3 4 5 6 7 9 X 9 0 =	careful!

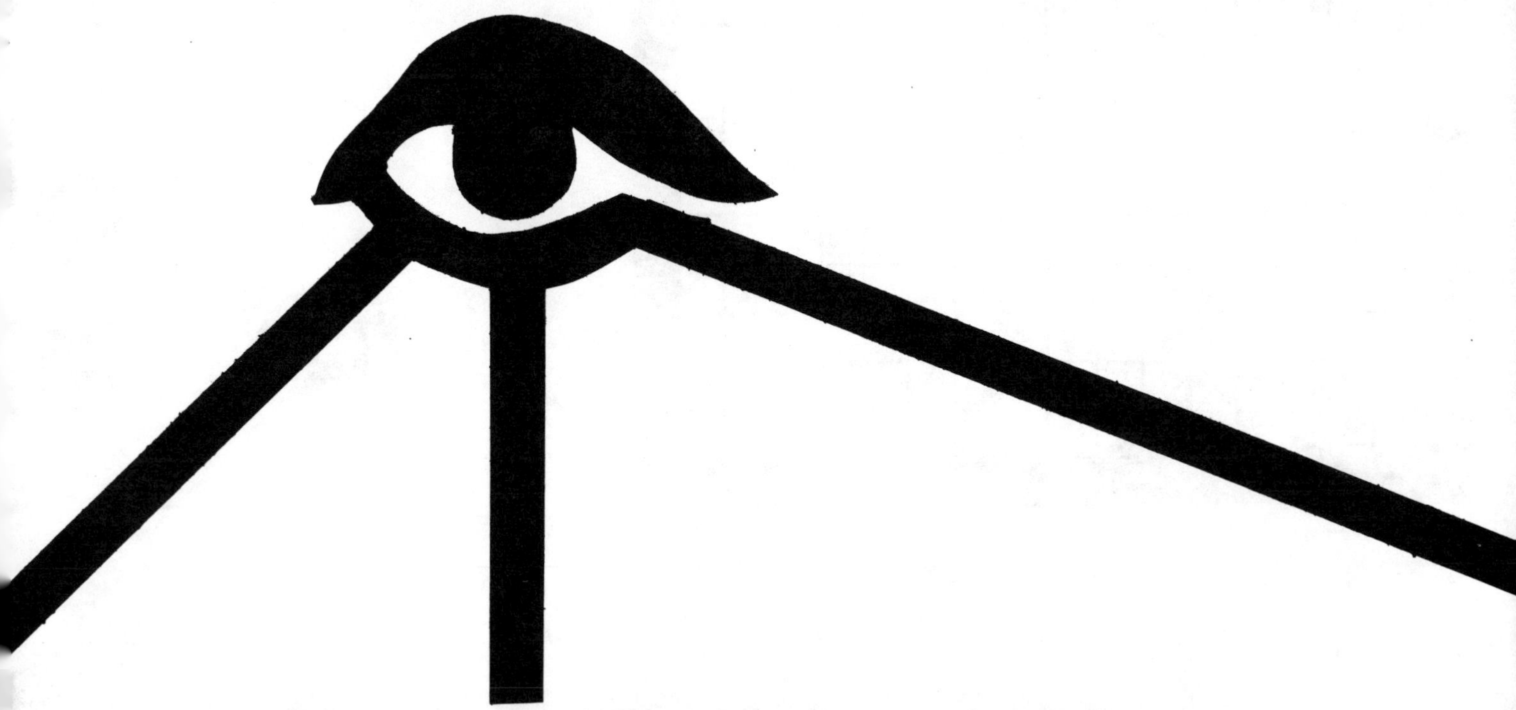

Magic* Numbers II.

name ______________________

period ______________________

date ______________________

3 X 37 = ___

6 X 37 = ___

9 X 37 = ___

12 X 37 = ___

15 X 37 = ___

18 X 37 = ___

21 X 37 = ___

24 X 37 = ___

27 X 37 = ___

careful! 30 X 37 = ____

name ____________

period ____________

date ____________

Magic Pyramid 1.

9 X 9 + 7 = ____

98 X 9 + 6 = ____

987 X 9 + 5 = ____

9876 X 9 + 4 = ____

98765 X 9 + 3 = ____

987654 X 9 + 2 = ____

9876543 X 9 + 1 = ____

98765432 X 9 + 0 = ____

name ______________________

period ______________________

date ______________________

Magic Pyramid 2.

9 X 9 = ☐☐

79 X 90 = ☐☐☐☐

679 X 900 = ☐☐☐☐☐☐

5679 X 9000 = ☐☐☐☐☐☐☐☐

45679 X 90000 = ☐☐☐☐☐☐☐☐☐☐

345679 X 900000 = ☐☐☐☐☐☐☐☐☐☐☐☐

2345679 X 9000000 = ☐☐☐☐☐☐☐☐☐☐☐☐☐☐

name ______________________

period ______________________

date ______________________

Magic* Pyramid* 3.

7 × 9 =

77 × 99 =

777 × 999 =

7777 × 9999 =

77777 × 99999 =

777777 × 999999 =

7777777 × 9999999 =

name

period

date

Magic* Pyramid* 4.

9 × 9 =

99 × 99 =

999 × 999 =

9999 × 9999 =

99999 × 99999 =

999999 × 999999 =

9999999 × 9999999 =

Magic* Pyramid* 5.

name ______________________

period ______________________

date ______________________

1 × 8 + 1 =

12 × 8 + 2 =

123 × 8 + 3 =

1234 × 8 + 4 =

12345 × 8 + 5 =

123456 × 8 + 6 =

1234567 × 8 + 7 =

12345678 × 8 + 8 =

123456789 × 8 + 9 =

name

period

date

Magic Pyramid 6.

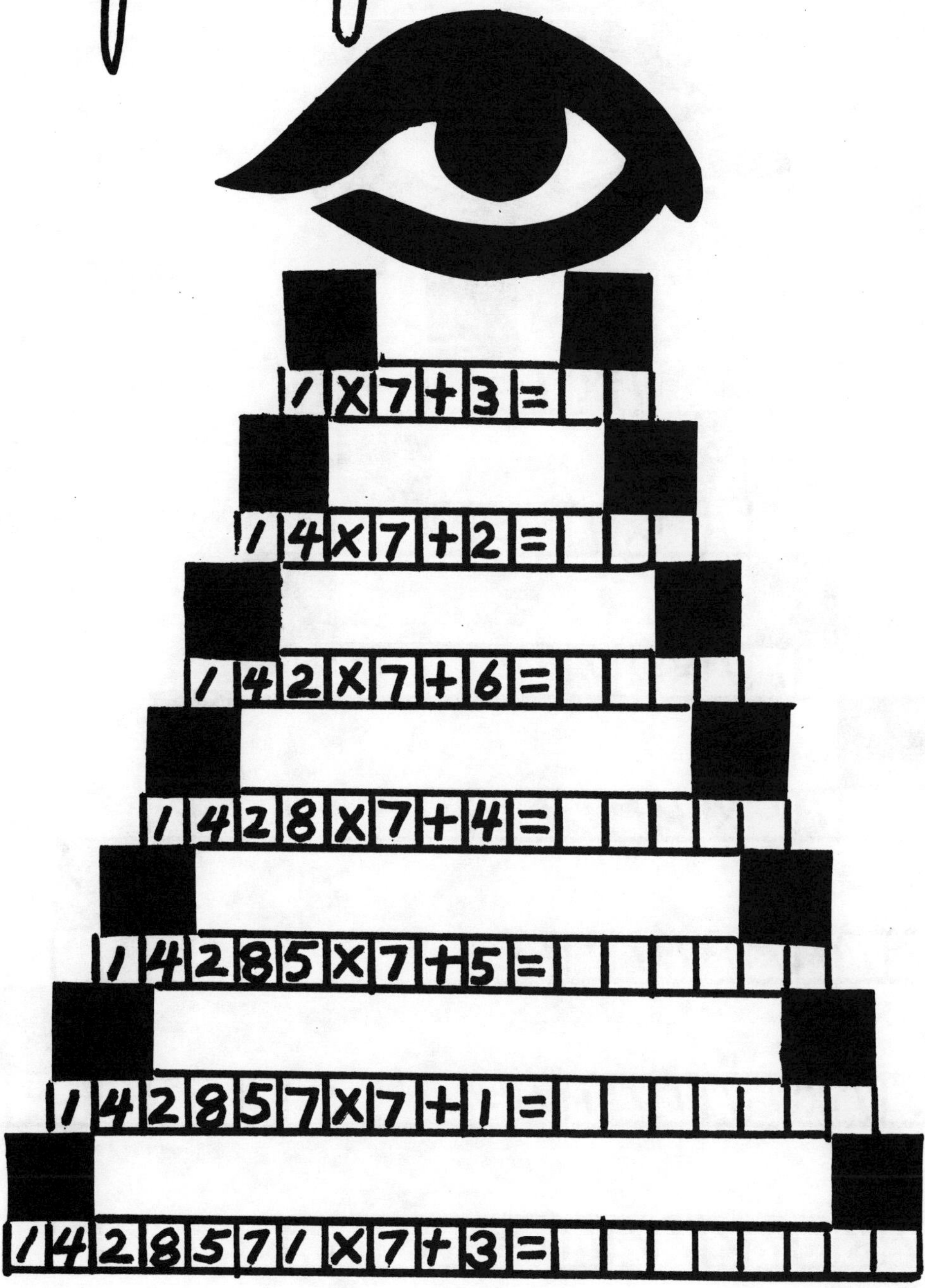

name ______________

period ______________

date ______________

Magic Pyramid 7.

name ______________

period ______________

date ______________

Magic* Pyramid* 8.

4 × 4 =

34 × 34 =

334 × 334 =

3334 × 3334 =

33334 × 33334 =

333334 × 333334 =

3333334 × 3333334 =

name ____________________

period ____________________

date ____________________

Magic* Pyramid* 9.

6 × 7 = ____

66 × 67 = ____

666 × 667 = ____

6666 × 6667 = ____

66666 × 66667 = ____

666666 × 666667 = ____

6666666 × 6666667 = ____

name ______________________

period ____________________

date ______________________

Magic* Pyramid* 10.

7 X 7 = ____

67 X 67 = ____

667 X 667 = ____

6667 X 6667 = ____

66667 X 66667 = ____

666667 X 666667 = ____

6666667 X 6666667 = ____

name ____________________

period ____________________

date ____________________

Magic* Pyramid* 11.

1 X 1 =

10 X 10 =

100 X 100 =

1000 X 1000 =

10000 X 10000 =

100000 X 100000 =

1000000 X 1000000 =

HINTS AND ANSWERS

1 Magic Squares-1

Square # 1

8 1 6
3 5 7
4 9 2

Square # 2

72 9 54
27 45 63
36 81 18

Square # 3

10 2 8
5 7 9
6 11 4

2 Magic Squares-2

Magic sum is 15
Magic sum is 2310
Magic sum is 1845

3 Magic Squares-3

17	24	1	8	15
23	5	7	14	16
4	6	13	20	22
10	12	19	21	3
11	18	25	2	9

4 Magic Squares-4

30	38	47	1	10	19	28
37	46	7	9	18	27	29
45	6	8	17	26	35	36
5	14	16	25	34	42	44
13	15	24	33	41	43	4
21	23	32	40	49	3	12
22	31	39	48	2	11	20

5 Multiplicative Magic Squares-1

12 1 18 216
9 6 4 216
2 36 3 216, etc.

$2^2(3)$ 1 $2(3)^2$
3^2 $2(3)$ 2^2
2 $2^2(3^2)$ 3

6 Multiplicative Magic Square-2

45 1 75 3375, etc.
25 15 9
3 225 5

7 Multiplicative Magic Square-3

20 1 50 28 1 98
25 10 4 49 14 4
2 100 5 2 196 7

8 Multiplicative Magic Square-4

63 1 147 99 1 363
49 21 9 121 33 9
3 441 5 3 1089 11

9 Magic Wheels and Triangles

7
6 8
1 5 9
2 4
3

5
7 3
6 4
2 9 1 8

2
5 4
9 8
1 6 7 3

Variations are likely

10 Magic Stars

1
4 14 10 2
12 13
3 9
5 11 8 6
7

10
4 7 9 6
8 5
1 11 12 2
3

1
2 11 16 5
12 9
10 8
6 7
14 3 4 13
15

11 Magic Maltese Multiplication

Center 6 Arm 1 12 18
12 2 9
Underlined 2 3 36
LARGE circles 4 18 3

12 Magic Roman Addition Cross

ARM II
I XIX
I+X

I+X
I I
XX

ARM I
XX III
XI

IX
XIX II
I

13 Magic Patterns-1

1. D 3. H 5. G 7. C 9. B
2. A 4. I 6. E 8. F

14 Magic Patterns-2

1. 111,111,111 E
2. H
3. F
4. C
5. I
6. G
7. B
8. D
9. A

15 Magic Patterns-3

1. 111,111 E 4. B 7. A
2. F 5. H 8. G
3. I 6. C 9. D

16 Magic Patterns-4

1. 111 C 4. H 7. G
2. E 5. F 8. I
3. A 6. D 9. B

17 Magic Patterns-5

(0) .012012... A (5) .567567... G
(1) .123123... C (6) .678678... F
(2) .234234... E (7) .789789... D
(3) .345345... H (8) .89.89.... B
(4) .456456... I

18 Magic Powers-1

10, 100, 1000, etc.

19 Magic Powers-2

153	6^3
370	9^3
371	12^3
407	10^3

20 Magic Powers-3

5	61
13	65
17	65
25	73
29	85
37	85
41	89
53	91

21 Magic Powers-4

9 $88 + 209 = 297$, 297^2
45 $494 + 209 = 703$, 703^2
55 $998 + 001 = 999$, 999^2
99

22 Magic Operations-1

3/2	16/5	49/8
4/3	25/6	
9/4	36/7	

23 Magic Operations-2

2, 3, 4, 5, 6, 7, 8

24 Magic Operations-3

9/2	25/4	49/6	81/9
16/3	36/5	64/7	

25 Magic Operations-4

3/2	5/4	7/6
4/3	6/5	8/7

26 Magic Matrix-1

Magic sum is 36

28 Magic Matrix-2

Magic sum is 0

30 Magic Matrix-3

Magic sum is 48

31 Magic Matrix-4

Answers will vary

32 Magic Mirrors-1

81	18
72	27
49	94
499	994
144	441
169	961
10404	40401
10609	90601
12544	44521
12769	96721
14884	48841

33 Magic Mirrors-2

510, 810, 540, 8910, etc.

34 Magic Mirrors-3

1236544	4456321
1238769	9678321
1256641	1466521
1468944	4498641
123476544	445674321
123498769	967894321
123676641	146676321
123698884	488896321
23456965432	23456965432
13457975431	13457975431
12367886521	12568876321

35 Magic Numbers-1

09109
18218
27327
36436
45545
54654
63763
72872
81981
91090

36 Magic Numbers-2

1999998
2999997
3999996
etc
8999991
9999990

37 Magic Numbers-3

321489
729316
328059
3285697041
3719048256
1492536780
1952873460
1537496280
9231874560

38 Magic Numbers-4

999999999
1999999998
2999999997
etc
8999999991
9999999990

39 Magic Numbers-5

7559 Need we say more?

40 Magic Numbers-6

111111
222222
etc
999999
1111110
1222221
1333332
1444443
etc
1999998

HINTS AND ANSWERS-3

41 Magic Numbers-7

1 1 1 1 1 1

2 2 2 2 2 2

3 3 3 3 3 3

etc

9 9 9 9 9 9

42 Magic Numbers-8

5796

5796

5346

5346

7254

7632

4396

6952

7852

43 Magic Numbers-9

1 1 1 1 1 1

2 2 2 2 2 2

etc

9 9 9 9 9 9

1 1 1 1 1 1 0

1 2 2 2 2 2 1

1 3 3 3 3 3 2

etc

1 9 9 9 9 9 8

2 1 1 1 1 0 9

2 2 2 2 2 2 0

2 3 3 3 3 3 1

2 4 4 4 4 4 2

etc

2 9 9 9 9 9 7
3 1 1 1 1 0 8

3 2 2 2 2 1 9

3 3 3 3 3 2 0

44 Magic Numbers-10

1 1 1 1 1 1 1 1 1

2 2 2 2 2 2 2 2 2
etc

9 9 9 9 9 9 9 9 9

1 1 1 1 1 1 1 1 1 0

45 Magic Numbers-11

1 1 1, 2 2 2, 3 3 3

etc to 1 1 1 0

46 Magic Pyramid-1

8 8
8 8 8
8 8 8 8 etc

47 Magic Pyramid-2

8 1

7 11 0

6 111 00

5 1111 000
etc

48 Magic Pyramid-3

6 3

7 62 3

77 622 3

777 6 222 3
etc

49 Magic Pyramid-4

8 1
9 8 0 1
99 8 00 1
9998 000 1
etc

50 Magic Pyramid-5

9
98
987
9876
etc

51 Magic Pyramid-6

10
100
1000 etc

52 Magic Pyramid-7

1
1 2 1

1 232 1

1234 321
etc

53 Magic Pyramid-8

1 6

11 5 6

111 55 6

1111 555 6
etc

54 Magic Pyramid-9

4 2
44 22
444 222
etc

55 Magic Pyramid-10

4 9

44 99

444 999
etc

56 Magic Pyramid-11

1

100

1 00 00

etc

ACTIVITY RESOURCES PUBLICATIONS

RESOURCE BOOKS

The Fabric of Mathematics
The Tapestry of Mathematics
The Correlation of Math & Science
Hands-On Math for Secondary

MANIPULATIVES

Algebra in the Concrete
Attribute Acrobatics
Base Ten Mathematics
Coin Stamp Mathematics
Colored Cube Activities
Discover It!
Focus on Calculator Math
Fraction Tile Program
Geoblocks and Geojackets
It's a Tangram World
Let's Pattern Block It
Math Balance Book
Metric Multibase Math
Multilinks Explorations
Number Triangles
People Piece Primer
People Piece Puzzles
Skateboard Practice, +,-
Skateboard Practice, X,/
Solid Sense in Mathematics
Tangram Geometry in Metric

METRIC BOOKS

Do Decimals First
Making Metric Maneuvers
Metrics All Around
Metric System of Measurement
Merry Metric Cookbook

GAMES

Enhance Chance
Friendly Games to Make
Games and Aids for Math
Place Value Games
Ready to Play, X, /
Tricks of the Trade
Tripletts

ACTIVITY BOOKS

Brain Busting Puzzles
Calculators and Magic
DeCode Puzzles
Focus on Decimals, 1,2
Focus on Fractions, 1,2,3
Focus on Geometry
Focus on Pre-Algebra
Graphiti, 1,2
Holiday Mathematics
Horoscope Shade-Ins
Key Word Problems
Laboratory Laughter
Puzzling into Algebra
Puzzling into Geometry
Skeletons, Word Problems
Time and Time Again

ART AND MATHEMATICS

Bucky for Beginners
Create a Cube
Mathmagic with Flexagons
Math and Art
Maze Book
Pholdit
Puzzles in Space

ACTIVITY RESOURCES COMPANY, INC.
P.O. BOX 4875
HAYWARD, CALIFORNIA 94540